– MESMERISM –

THE DISCOVERY OF ANIMAL MAGNETISM

Franz Anton Mesmer

Translated from the French

by

Joseph Bouleur

HOLMES PUBLISHING GROUP

MESMERISM — The Discovery of Animal Magnetism (1779)–Copyright © 1998, 2006, 2007, 2009 by Holmes Publishing Group LLC and Joseph Bouleur and his designee, Université de Lyon. New material © 2006, 2007 by Joseph Bouleur and his designee, Université de Lyon. All rights reserved. No part of this book may be reproduced or utilized in any form or by any means, electronic or mechanical, including photocopying, recording or by any information storage and retrieval system, without permission in writing from Holmes Publishing Group LLC except in the case of brief quotations embodied in critical articles and reviews.

ISBN 1-55818-382-5

FIFTH EDITION,

REVISED

— 2009 —

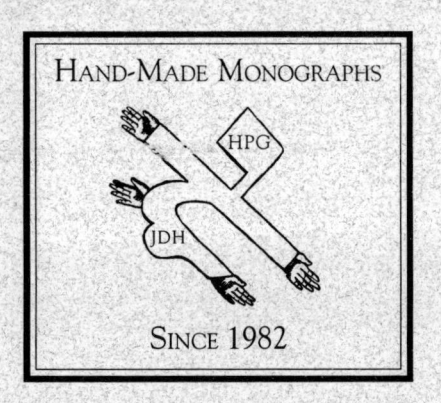

HAND-MADE MONOGRAPHS
HPG
JDH
SINCE 1982

HOLMES PUBLISHING GROUP LLC
POSTAL BOX 2370
SEQUIM WA 98382 USA
JDHOLMES.COM

Foreword

The discovery, so long attempted, of a principle acting on the nervous system should be of great interest to all. It offers a double significance. First, as an increase in self-knowledge, and secondly, this discovery leads to greater happiness as it affords a means of curing the ills of the body which have heretofore been corrected with little success. The qualities and the special nature of this system of healing were responsible, some years ago, for the popular support given to the possibility of real hope in difficult cases. It is by the perversion of my thoughts and methods that envy, presumption and incredulity have in a brief period succeeded in relegating my methods to the status of illusions. This has resulted in their fall into oblivion.

I have hoped (and vainly) to revive them by much substantive evidence, but prejudice and bigotry has triumphed. Truth has been sacrificed. But, it will be wondered, what is the content of this new method? How and where did you find it? What advantages may we assume? Why have you not used it to enrich your fellow citizens? All of these questions have been put to me since my arrival in Paris.

I wish to answer clearly in order to not only provide a general idea of the system I propose, but to also free it from the errors which has encircled the concept and expose the vicissitudes which form the obstacles to its being made generally known. This Dissertation serves that purpose, but it is merely a forerunner.

I shall later impart more knowledge regarding the practical rules of the method itself.

Within this light, I ask sincerely that the reader consider this brief work. I am aware that my thesis will present difficulties, but these difficulties cannot be overcome by reason alone, but must have the assistance of experience to be successful.

Experience alone will scatter the clouds and shed light on this important truth: Nature provides a universal means of healing and the preserving of the human being.

By nature, the human being is an observer. From birth, his primary purpose is to perceive impressions in order to learn the use of the senses. The eye would be of little use if Nature did not cause him to pay attention to the minute variations of which his observation is capable. It is by the alternating effect of satisfaction and deprivation that he "knows" of the existence of light in its different degrees and levels. The observer would remain ignorant of the distance, size and shape of objects if he did not ascertain these by comparing and combining the impressions of all of the other organs—even to understanding how to correct one by the other. Most impressions are therefore the result of his reflections on the sensations assembled in the sense organs. Accordingly, the human being spends the early years of life in the acquisition of a correct use of his senses. From Nature, the person acquires his gift of observation, a gift enabling him to form himself. The perfection of his faculties depends on its more or less constant application. But his best and fullest attention is reserved for those objects to which he has a particular attraction.

Observation of the impressions which Nature constantly produces on each individual is not just the special domain of Philosophers, but by the universality of the effects, it makes an observer of almost every individual. These observations, multiplied in every time and place, leave nothing to be desired as regards their reality.

The activity of the human mind, together with its unsatisfied ambition for knowledge, in seeking to perfect knowledge previously acquired, abandons observation, replacing it by vague and sometimes frivolous speculation. It creates and accumulates systems whose only merit is their mysterious abstraction. It loses sight of the truth, sometimes imperceptibly, to such an extent that it sets ignorance and superstition in its stead. Human knowledge, thus perverted, ceases to own the reality which it began with.

Now and again, Philosophy made efforts to free itself of so-called errors and prejudices, but by seeking to eliminate these by rancorous vigor, it has cloaked the ruins with too much disdain, forgetting the precious things hidden therein. We see among the different peoples the same prejudices preserved in a form so dishonorable for the human mind that it seems improbable that it was created for that purpose.

Imposture and aberration of reason has been used to win over nations and peoples to adopt such absurd and ridiculous systems as we see today. Much of this chicanery has been in vain. Truth alone and the general interest would have conferred their universal nature on these opinions, but these concepts remain false.

Nevertheless, it may be asserted that among the ignorant opinions of all ages, whose principles are not rooted in the human heart, there are but few which, however ridiculous they may appear, cannot be regarded as embodying a primal truth.

These are my reflections on knowledge in general, and more particularly on the fate of the doctrine of the influence of celestial bodies on the planet we inhabit. These reflections have induced me to seek, among the ruins of that science, brought so low by ignorance, what it might have contained that was useful and true.

In accordance with my ideas on this subject, I published at Vienna in 1766, a thesis on the influence of the planets on the human body. According to the familiar principles of universal attraction, ascertained by observations which teach us how the planets mutually affect one another in their orbits, how the sun and moon cause and control the ocean tides on our globe, and in the atmosphere, I asserted that those spheres also exert a direct action on all the parts that go to make up animate bodies, in particular on the nervous system, by an all-penetrating fluid. I denoted this action by the effects of intensification and remission of the properties of matter and organic bodies, such as gravity, cohesion, elasticity, irritability, electricity, and others.

I maintained that just as the alternate effects, in respect of gravity, produce in the sea the phenomenon which we term ebb and flow, so the intensification and remission of the said properties, being subject to the action of the same principle, cause in animate bodies alternating effects similar to those endured by the sea. By these considerations I established that the animal body, being subjected to the same action, likewise underwent a kind of ebb and flow. I supported this theory with different examples of periodic revolutions. I named the property of the animal body that renders it liable to the action of heavenly bodies and of the earth: Animal Magnetism. I explained by this magnetism the cyclical changes which we observe in sex, and in a general way, those changes which physicians have observed during the course of an illness.

I had no other ambition than to arouse the interest of physicians. I did not succeed. I soon became aware that I was being associated with folly; that I was being treated like a man with a fabricated system; and that my tendency to quit the customary path of Medicine was being construed as a crime. I have never concealed my thinking and opinion in this last respect. I am unable to convince myself that the art of healing has made the progress of which we incessantly boast.

Indeed, I held that the further we advanced in our knowledge of the mechanism and the economy of the animal body, the more we were compelled to admit our continuing insufficiency. The knowledge that we have gained today about the nature and action of the nerves, imperfect though it is, leaves no doubt in this respect. We know that the nerves are the principal agents of sensation and movement, but we are unable to restore them to their natural order when there has been a general imbalance. We confess this to our shame. The ignorance of bygone centuries on this point has sheltered physicians. The superstitious confidence which physicians assumed and which inspired their specifics and formulae made them despotic and presumptuous. They remain so today.

I have too much respect for Nature. I cannot convince myself that the individual preservation of the human being has been left to a mere chance of discovery and to the vague observations that have been made in the course of a number of centuries, which finally became the domain of the few.

Nature has provided everything for the existence of the individual. Propagation takes place without "system" and without trickery. Why should preservation be deprived of the same advantage? The preservation of animals affords proof that the contrary is the case.

A non-magnetized needle, when set in motion, will only take a determined direction by chance or hazard. A magnetized needle will, after various oscillations proportional to the impulse and magnetism received, will recover its initial position and remain there. Thus the harmony of organic bodies, when once interfered with, goes through the uncertainties of my first hypothesis, unless it is brought back and determined by the Universal Agent, whose existence I recognize:—it alone can restore harmony in the natural state.

Thus we have seen, in all ages, maladies which become worse or better, with or without the help of Medicine, in accordance with different systems, and by the most discordant methods. These considerations convince me that there exists in Nature one universally acting principle which, independently of ourselves, operates in and through what we vaguely attribute to Art, Chance and Nature.

I have subjected my ideas to experience for more than twelve years. I have devoted myself to the most accurate observations of all types of disease. I have had the satisfaction of seeing the maxims I projected borne out in instance after instance.

In the years 1773 and 1774, I undertook the treatment of a young lady. She was twenty-nine at the time and was named Oesterline. For several years, she had been subject to a convulsive malady, the most troublesome symptoms being that the blood rushed to her head and brought about a most cruel toothache as well as an extreme pain in the ears. This was followed by delirium, rage, vomiting and fainting. This offered a highly favorable circumstance to me for observing accurately the ebb and flow to which Animal Magnetism subjects the human body. [For more details on this case and others to follow, see Swainson & Clary's *Franz Anton Mesmer*, Holmes Publishing Group, 1999.] The patient often had beneficial periods, followed by a remarkable degree of alleviation; however, the comfort was always momentary and imperfect.

The desire to ascertain the cause of this imperfection, and my own uninterrupted observations, brought me again and again to a point of recognizing Nature's handiwork, and of penetrating it sufficiently to forecast and affirm, without hesitation, the different stages of the illness. Encouraged by this first success, I no longer held any doubts as to the possibility of healing, if I were able to discover the existence, among the substances of which our globe is made, of an action that is also reciprocal and similar to that of the heavenly bodies, by means of which I could imitate artificially the periodic revolutions of the ebb and flow just referred to.

I had some previous knowledge of the magnet and its powers: its operation on iron, the ability of our body fluids to receive that mineral. The various tests previously carried out in France, Germany and Britain for stomach ache and toothache were known to me. These reasons, together with the analogy between the properties of this substance and the general system, caused me to regard it as being sound for this type of test. To assure the success of this test, in the interval of the attacks, I prepared the patient by the continuous use of chalybeates.

My social relations with Father Hell, Jesuit and Professor of Astronomy at Vienna, gave me an opportunity of asking him to have made for me a number of magnetized pieces by his craftsman. These needed to be of a convenient shape for application. He was kind enough to do this for me and to let me have them.

On 28 July, 1774, after the patient had a renewal of her attacks, I applied three magnetized pieces to the stomach and both legs. In a few moments, this application was followed by extraordinary sensations. She felt some painful currents internally of a subtle material which, after different attempts at taking a direction, made their way toward the lower part of her body. The symptoms of her illness ceased for more than six hours. However, the following day, the patient's condition made it necessary for me to carry out the same test again, but agin, I obtained the same success as previously.

My observation of these effects, coupled with my ideas on the general system, provided me with fresh information. While confirming previous ideas about the influence of the *Universal Agent*, it convinced me that another principle was causing the magnet to act as well — the magnet itself being incapable of such action on the nerves. I understood that I was drawing closer to the Imitative Theory, which composed the subject of my research.

Some days later, I met with Father Hell and mentioned to him in the course of conversation that the patient was in a better state of health. I elaborated on the good effects of my process and my hopes, on the strength of this operation, of soon discovering a means of curing those who suffered from nerve disorders.

Not long afterwards, I became cognizant from public sources, that this man of religion, misapplying his fame in astronomy and wishing to appropriate for himself a discovery of whose nature and benefits he was entirely ignorant, had taken upon himself to publish the following — that by means of some magnetized pieces, to which he attributed a special virtue depending on their shape, he had obtained the means of curing the most difficult nerve disorders. To bolster this opinion, he sent to a number of Universities some sets consisting of magnetized pieces of all shapes, with an outline their use with various maladies. See following the words he himself expressed: "I have discovered in these shapes, which agree with the magnetic vortex, a harmony on which depends their specific virtue in cases of illness; it is owing to the lack of this knowledge that the tests conducted in England and France met without success." He finished by saying, that he had communicated everything to the physicians, and to me in particular, and would continue to experiment with the method.

The writings of Father Hell on this subject energized the public mind with the unsound opinion that the discovery in question consisted in the mere use of the magnet. I refuted this error, by verifying the existence of *Animal Magnetism*, as a force distinct from the Magnet. The public held to its erroneous belief solely on the basis of having first received notice from a man of good reputation. Regardless, I continued my experiments with different disorders so as to increase my knowledge and perfect the method of it.

I knew Baron de Stoerck, President of the Faculty of Medicine at Vienna and Chief Physician to Her Majesty. Since I knew him quite well, I decided to acquaint him with the nature of my discovery and its purpose. For this purpose, I placed before him the details of my operations, particularly as regards the communication and currents of animal magnetic matter. I invited him to verify the efficacy of the treatment for himself, stating that it was my intention to report to him in future all progress that I might make in this new science. To show good faith, I proceeded to detail my methods to him without reservation.

The natural fear of this physician, and based on questionable motives, caused him to reply that he would have nothing to do with my discoveries. Further, he begged me not to compromise the faculty by linking it with an innovation of this kind.

The prejudice of the public, a prejudice based on uncertainty about my methods, caused me to publish, on January 5, 1775, a *Letter to a Foreign Physician*. In this writing, I gave an exact idea of my theory, a chronicle of the success obtained and the triumph I had reason to hope for. I undraped the nature and action of ANIMAL MAGNETISM, giving the analogy between its properties and those of the magnet and electricity. I added, "All bodies were, like the magnet, capable of communicating this magnetic principle; that this fluid penetrated everything and could be stored up and concentrated, like the electric fluid; that it acted at a distance; that animate bodies were divided into two classes, one being susceptible to this magnetism and the other to an opposite quality that suppresses its action." Finally, I accounted for the various sensations and based these assertions on experiments which enabled me to put them forward.

A few days prior to the publication of this *Letter*, I was told that a Mr. Ingenhousze, a member of the Royal Academy of London, by entertaining the nobility and other distinguished personages with experiments in electricity, and particularly by the skill with which he used the magnet, had acquired the reputation of being a physician. I heard, as I said, that when this gentleman learned of my experiments, he treated them as the most vain of imaginings. He went so far as to say that only the English genius was capable of such a discovery, if it could be done at all. Mr. Ingenhousze came to see me, but not to get further information. No, his sole intention was to persuade me that I was opening myself to ridicule and that I should suppress all publicity immediately. I replied that I thought him not sufficiently talented to give advice, and moreover, that I would take great pleasure in convincing him of my views at the first opportunity. This came two days later.

Miss Oesterline contracted a chill, causing a sudden stoppage, and relapsed into her former spells. I invited Mr. Ingenhousze to come around. He came in the company of a young physician. The patient by this time had fallen into a fainting fit with convulsions. It was indeed a most favorable opportunity for convincing the gentleman of the existence of the principle I had announced, and the property which it substantiates. I asked him to approach the patient, while I withdrew from her, instructing him to touch her. She made no movement or sound. I called him back to me, and transferred animal magnetism to him by taking his hands. I then had him approach the patient once more. I kept at a distance, telling him to touch her a second time which resulted in convulsive movements. I had him repeat this touching process a number of times, which he did with the tip of his finger, changing the direction each time. In every case, much to his great astonishment, a convulsive effect brought about in the part touched.

When this session was over, he told me he was convinced. I suggested a second visit. We withdrew from the patient and I offered Mr. Ingenhousze six china cups. I asked him to point to the one which he wished me to transfer the magnetic quality. I operated upon the cup of his choice, and then applied the six cups to the patient's hand in succession; on reaching the cup that I had touched, the hand suddenly made a motion and gave signs of pain. Mr. Ingenhousze obtained the same result when he applied the six cups in a similar manner.

I had these cups returned to their place. After a certain interval and holding him by one hand, I asked him to touch with the other hand any cup he wished. He did so, the cups were brought to the patient, as before, and yielded the same result.

The communicability of the principle was now well-established in Mr. Ingenhousze's eyes, and I suggested a third experiment to show its action at a distance and its penetrating quality. I pointed my finger at the patient at a distance of eight paces. In the next instant, her body convulsed to the point of raising her body on her bed with every appearance of pain. I continued, in the same position, to point my finger at the patient, placing Mr. Ingenhousze between herself and me. She experienced the same sensations.

Having repeated these tests to Mr. Ingenhousze's satisfaction, I asked him if he was convinced of the marvelous properties about which I had told him. I even offered to repeat the entire proceedings if he were not. His reply that he needed nothing further. He was convinced, but owing to his friendship with me, he appealed to me not to make any public statement on this subject, so as not to draw the public's incredulity. We parted at this point. I went back to the patient to continue the treatment. It was a most successful session, and that same evening, I restored her to the normal course of nature, thereby putting an end to the pain brought about by the stoppage.

Two days later I was astonished to hear that Mr. Ingenhousze was making statements in public that were quite the reverse of his utterances in my house, and was denying the success of the different experiments he had witnessed. He effected to confuse ANIMAL MAGNETISM with the magnet and was endeavoring to damage my reputation by spreading the report that with the aid of a number of magnets which he had brought with him, he had succeeded in unmasking me, proving that it was nothing but a ridiculous fraud.

This seemed to me unbelievable; I had some difficulty in bringing myself to regard Mr. Ingenhousze as their author. However, his association with the Jesuit Hell, and the latter's irresponsible writings in support of such odious insinuations, aimed at ruining the effect of my Letter of January 5, 1775, cast all doubt from my mind that Mr. Ingenhousze was indeed guilty. I refuted Father Hell and was about to have an indictment drawn when Miss Oesterline, who had been informed of Mr. Ingenhousze's procedure, was so affronted at finding herself so compromised that she relapsed into her former state, which was aggravated by a nervous fever.

Miss Oesterline's illness claimed the whole of my attention for a fortnight. In these circumstances, by continuing my research, I was fortunate in overcoming the difficulties which stood in the way of my progress and of giving my theory the desired perfection. The cure of this young lady represented the first fruits of my success, and I had the satisfaction of seeing her from this time on in excellent health. She has since married and had some children as well.

During this fortnight, I was determined to justify my conduct and to give the public a correct idea of my abilities by unmasking Mr. Ingenhousze's conduct. I informed Mr. de Stoerck, asking him to obtain orders from the Court for a Faculty Commission to be acquainted with the facts, so that it might verify my work, and make it known to the public. This step was agreeable to the physician, for he seemed to share my viewpoint and promised to act accordingly, remarking, however, that he would not serve on the Commission.

Several times, I suggested that he should visit Miss Oesterline and satisfy himself as to the success of my treatment. His replies were always vague and uncertain. I expressed to him how it would benefit humanity to have my method adopted by the hospitals, and asked him to demonstrate its utility as soon as possible at the Spanish Hospital. He agreed and gave the necessary instructions to Dr. Reinlein, a physician at the hospital.

Dr. Reinlein was a witness to the effects and general usefulness of my visits over a period of eight days. On several occasions he expressed surprise and reported to Mr. de Stoerck. However, I shortly became aware that different impressions had been addressed to this leading physician. I saw him almost every day and continued to insist on the institution of a Commission, reminding him of the interesting matters about which I had told him. I saw nothing in him, but an indifference and coldness in his attitude, whenever the topic was raised.

I could not obtain satisfaction about the commission and, moreover, Mr. Reinlein ceased reporting to me (I found out that this change of conduct was due to Mr. Ingenhousze). I realized at this time that I could not staunch the flow of intrigue and sought the shelter of silence, a poor consolation.

Becoming brazen by reason of the success of his plans, Mr. Ingenhousze acquired fresh vigor. He crowed his incredulity, and it wasn't long before he classed those of independent judgment, or who did not share his views, as feebleminded. It will be understood that this tactic was quite enough to alienate the public. In fact, I began to be conceived as a visionary, especially as the slight by the Faculty appeared to support that opinion.

In the following year, the very same opinion was shared by Dr. Klinkosch, a Professor of Medicine at Prague, who, without having the slightest knowledge of me (and without the least idea of the true state of the matter), was sufficiently foolish (to use no stronger words), as to publish in a journal, the details of the frauds attributed to me by Mr. Ingenhousze.

I felt that truth could not find a better advance than in the facts regardless of public opinion. So I undertook the treatment of various disorders, including a hemiplegia, the result of apoplexy; stoppages, spitting of blood, frequent colics and convulsive sleep from childhood, combined with a spitting of blood and normal ophthalmia. Mr. Bauer, Professor of Mathematics at Vienna, was attacked by this latter malady. My work was crowned with a fine success, and Professor Bauer himself was honest to introduce into the public record a detailed report on his cure. However, prejudice once again had the upper hand, but this time, I had the good fortune of being quite well known to a great Minister, a Privy Councilor and an Aulic Councilor, friends of humanity, who had recognized truth for themselves, seeing that they upheld and protected it. They made several attempts to lighten the shadows in which it was being enshrouded, but met with little success. It was objected that only the opinion of doctors was capable of deciding such an issue. Their good will was thus confined to their offers to give my writings the necessary publicity in foreign lands.

It was through this channel that my explanatory *Letter* of January 5, 1775 was transmitted to the majority of the scientific institutions, and to a few scientists. Only the Berlin Academy, on 24th March of that year, made a written reply in which by confusing the properties of ANIMAL MAGNETISM which I had expounded with those of the magnet, (which I only spoke of as a conductor), the reply was full of errors — its final opinion being that I was the victim of illusion!

This Berlin Academy was not the only one to make the mistake of confusing ANIMAL MAGNETISM with that of mineral magnetism, although I have always stressed in my writings that the use of the magnet, however convenient, was always imperfect without the assistance of the former theory. The doctors with whom I have been in correspondence, or who have endeavored to find out my methods in order to "steal" my discovery, have begun to spread it about that the magnet was the only means I employed, or perhaps that I used electricity as well, because it was well known that I had used both in the past. Instead of realizing the truth I was expounding, they have concluded from the fact that they obtained no success from the use of these two agents that the cures announced by myself were imaginary, and that my theory was nothing, but a chimera. The desire to disprove such errors permanently, and to do justice to the truth, determined me to make no further use of electricity, or of the magnet for that matter, from 1776 onwards.

The poor reception given to my discovery and the dashed hopes it held out to me for the future made me resolve to undertake nothing in public at Vienna. I decided instead to travel to Swabia and Switzerland in order to add to my experience, preferring to arrive at the truth through facts. I had the satisfaction of making some striking cures in Swabia, and of operating in the hospitals, before the eyes of doctors from Berne and Zurich. They were left in no doubt as to the existence of ANIMAL MAGNETISM, and the usefulness of my theory. It was gratifying to see the errors ascribed to me by my opponents corrected.

Between the years 1774 and 1775, an ecclesiastic, who was a man of good faith, but of excessive zeal, was operating in the diocese of Ratisbon on various disorders of a nervous nature. He used means that appeared supernatural to the less informed in that district. His reputation extended to Vienna, where society was divided into two halves: one regarded his methods as imposture and fraud, while the other looked upon them as miracles performed by Divine power. I believed both were wrong. My experience at once told me that the man in question was merely a tool of Nature. He was blessed by his profession and assisted by fate in that these furnished him with certain natural talents enabling him to find out the periodical symptoms of maladies without knowing their cause. The end of such paroxysms was held to be a complete cure, and time alone could lift the veil from the public.

Returning to Vienna towards the end of 1775, I passed through Munich, where His Highness, the Elector of Bavaria, was kind enough to consult me on this subject, asking me whether

I could account for these pretended miracles. I carried out before his eyes some experiments that removed any prejudices he may have had and left him in no doubt as to the truth I announced. A short time later, the Scientific Institution of that city paid me the honor of admitting me as a member.

I returned to Bavaria for another visit in 1776. I secured similar success using my methods in the cure of illnesses of different kinds. In particular, I effected the cure of an imperfect *amaurosis*, accompanied by paralysis of the limbs, which was afflicting Mr. d'Osterwald, director of the Scientific Institution of Munich. He was generous to make public mention of this and the other results he had witnessed.

After returning to Vienna, I undertook no further work until the end of the year. I would not have changed my mind, but my friends were unanimous in opposing my decision. Their prodding, together with my desire to see truth prevail, aroused in me the hope of accomplishing this by means of new successes, particularly through some striking cure. With this end in view, I undertook the treatment of Miss Paradis, an eighteen year old girl, whose parents were well known; she herself was known to Her Majesty, the Queen-Empress, through whose generosity she received a pension. She had been blind since the age of four. It was a perfect *amaurosis*, with convulsions in the eyes. She was also a prey to melancholia, accompanied by stoppages in the spleen and liver. These stoppages produced fits of delirium and rage and she had become convinced that she was out of her mind.

I began to also treat Miss Zwelferine, a girl of nineteen years, who had been blind since the age of two owing to an *amaurosis*. This was accompanied by a very thick, wrinkled albugo with atrophy of the ball. She was afflicted with periodic spitting of blood as well. I found this girl in the Vienna orphanage, and her blindness was attested to by the Governors.

At the same time, I treated Miss Ossine, aged eighteen, who was in receipt of a pension from Her Majesty, as being the daughter of an officer in her armies. Her malady consisted of purulent phthisis and irritable melancholia, accompanied by fits, rage, vomiting, spitting of blood and fainting. These three patients and others besides were accommodated in my house so that I might continue my treatment without interruption. I was fortunate in being able to cure all three.

The parents of Miss Paradis, who witnessed her cure, and the progress she was making in the use of her eyesight, hastened to make this occurrence known, and how pleased they were. Crowds came to my house to make certain for themselves, and each one, after putting the patient to some kind of test, withdrew greatly astonished, with the most flattering remarks to myself.

The two Presidents of the Faculty, at the head of a deputation, came to see me at the repeated request of Mr. Paradis. After examining the young lady, they justly added their tribute. Dr. de Stoerck, one of these gentlemen who knew this young person particularly well, having treated her for ten years without the slightest success, expressed to me his satisfaction at so interesting a cure, and his regret at having so far deferred his acknowledgment of the importance of my discovery. A number of physicians, each for himself, followed the example set by our leaders and paid the same tribute to truth.

After such authentic recognition, Mr. Paradis was kind enough to express his gratitude in his writings, which were promulgated throughout Europe. It was he who afterwards published the interesting details of his daughter's recovery in the newspapers.

Among the physicians who came to see me to satisfy their curiosity was Dr. Barth, professor of anatomy, of diseases of the eye, and a cataract specialist. He even admitted on two occasions that Miss Paradis was able to use her eyes. Nevertheless, this man's envy caused him to state publicly that the young lady could not see, and that he had satisfied himself that she could not. He maintained this on the fact that she did not know or confused the names of objects shown to her. He was answered from all quarters that he was confusing the necessary inability of those blind from birth (or at a very tender age) with the knowledge acquired by blind persons operated on for cataract. How, he was asked, is it that a man of your profession can be guilty of so obvious an error? His impudence, however, found an answer to everything by asserting the contrary. A thousand witnesses had given evidence of the cure! He alone held the opposite view, in which he was joined by Mr. Ingenhousze of whom I previously referred.

These two individuals, who were at first regarded as fanatics by sensible, honest folk, succeeded in mounting a scheme to withdraw Miss Paradis from my care, her eyes still being in an imperfect state, and made it impossible for her to be presented to Her Majesty, as was to have been the case. This inevitably lent credence to their assertion of imposture. To bolster their case, they sought to influence Mr. Paradis, who began to be concerned that his daughter's pension, and several other advantages might be stopped. He consequently asked for his daughter's return. The girl, supported by her mother, showed unwillingness and fear in case the cure might be imperfect. The father insisted, and this dispute brought on her fits again and led to an unfortunate relapse. However, this had no effect on her eyes, and she continued to improve the use of them.

When her father saw she was better, but still egged on by the conspirators, he returned to the charge. He demanded his daughter with some anger and compelled his wife to do likewise. The girl resisted for the reasons cited before. Her mother, who had always supported her, and apologized for the lengths to which her husband had gone, came to tell me on 29 April, 1777 that she intended to remove her daughter instantly. I replied that she was free to do so, but if fresh accidents were the result, she could not count on my help.

These words were overheard by the girl, who was so overcome that she went into a fit. She was assisted by Count de Pellegrini, one of my patients. Her mother, who heard her cries, left me abruptly and seized her daughter angrily from the hands of the person who was assisting her, saying: "Wretched girl, you too are hand in glove with the people of this house!" as she flung her against the wall.

Immediately, the troubles of that unfortunate girl resumed. I hastened towards her to give assistance, but her mother, still livid with rage, threw herself on me to prevent me from doing so, all the time shouting insults at me. I had the mother removed by some members of my household and went up to the girl to assist her. While I was tending her, I heard more angry shouts, and there were repeated attempts to open the door of the room where we were.

It was Mr. Paradis, who having been warned by one of his wife's servants, was now invading my house with drawn sword. His intention was to enter the room where I was, while my servant was trying to remove him by guarding the door. The madman was at last disarmed, and he left my house breathing abuses on myself and my household.

Meanwhile, his wife had fainted. She was given attention and left some hours later. The unhappy girl was suffering from attacks of vomiting, fits and rages, which the slightest noise, especially the sound of bells, accentuated. She had relapsed into her previous blind state through the violence of the blow given her by her mother. I had some fears for the state of her brain as well. Oh, the sinister effects of that painful scene.

It would have been simple for me to take the matter to court, on the evidence of Count de Pellegrini and eight persons who were with me, to say nothing of other neighbors who could have acted as witnesses. I was solely concerned with saving Miss Paradis at this point and decided not to resort to legal redress. My friends protested in vain, pointing out the ingratitude exhibited by her family, and the wasted expenditure of my labors. I stayed to my first decision, and would have been content to overcome the enemies of truth and of my peace of mind, by good deeds.

The following day I heard that Mr. Paradis, in an endeavor to cover up his behavior, was spreading about the most evil insinuations regarding myself, always with a view to removing his daughter and proving, by her condition, the dangerous nature of my methods. I received through Dr. Ost, the Court physician, a written order from Dr. de Stoerck, in his capacity as head physician, dated Schoenbrunn, May 2, 1777, who called upon me "to put an end to the imposture" — (his own expression) — "and restore Miss Paradis to her family, if I thought this could be done without risk."

Who would have believed that Mr. de Stoerck, being so well informed by the same physician of all that had taken place in my house and, after his first visit, having come twice to convince himself of the patient's progress and the success of my methods, could have taken upon himself to use such offensive and contemptuous language to me? I had reason to expect the contrary from him. He should have been the defender of truth, not its enemy. I would even go so far as to say that as the repository of Her Majesty's confidence, one of his first duties under these circumstances should have been to protect a member of the Faculty whom he knew to be blameless, and to whom he had often given assurances of his esteem. My answer to this irresponsible order was that the patient could not be moved without running the risk of death.

Miss Paradis's critical condition no doubt made an impression on her father, and caused him to reflect. Through the intermediary of two reputable persons, he begged me to continue attending his daughter. I told him that I would do so only on the condition that neither he or his wife would ever again appear in my house. My treatment exceeded my best hopes. Nine days sufficed to calm down the fits entirely, and put an end to the disorders, but her blindness remained.

Fifteen days of treatment cured the blindness and restored the eyes to the condition known prior to the incident. To this period, I added a further two week's attention to improve and restore her health. The public then came to obtain proof of her recovery, and everybody gave me, even in writing, fresh evidence of satisfaction. Mr. Paradis, being assured of the good health enjoyed by his daughter through Dr. Ost, who, at his request and by my permission, followed the progress of the treatment, wrote a letter to my wife in which he thanked her for the care bestowed on his daughter. He wrote a note thanking me and apologizing for the past. He finished by asking me to send back his daughter so that she might enjoy the benefit of country air. He said that he would send her back to me whenever I might think necessary, so as to continue the treatment. I believed him in good faith. I returned his daughter to him on the 8th of June.

Scarcely a day passed before I heard that her family maintained that she was still blind and subject to fits. They showed her off and compelled her to imitate fits and blindness. This news evoked some contradictions by persons who were convinced to the contrary, but it was upheld and accredited by the obscure intriguers who used Mr. Paradis as their tool. I was unable to check its spread by the highest testimony. Therefore, in spite of my perseverance and my work, I have little by little seen relegated to the position of a conjecture, at the least as something uncertain, a truth that has been proven by experience.

It is not difficult to imagine how I might have been affected by the relentlessness of my enemies to do me harm and by the ingratitude of a family on which I had showered kindness. During the last half of 1777, I continued with the cure of Miss Ossine and the aforementioned Zwelferine whose eyes, it will be remembered, were in an even more serious condition than Miss Paradis's. I also persevered successfully with the treatment of my remaining patients, in particular Miss Wipior, aged nine, who had in one eye a growth on the cornea, known by the name of staphyloma.

I succeeded in removing the excrescence from the staphyloma to the extent that she was able to read sideways. There only remained a slight albugo in the center of the cornea, and I have no doubt that I would have caused it to disappear entirely, had circumstances permitted me to continue the treatment. However, being tired by my labors extending over twelve consecutive years, and still more by the continued animosity of my adversaries, without having reaped from my research and efforts any satisfaction other than that of which adversity could not deprive me, I felt that I had done my duty for my fellow citizens.

With the conviction that justice would one day be done me, I decided to travel for the sole purpose of securing the relaxation I was in such need of. However, to guard against insinuations as far as possible, I arranged matters so as to leave at home in my absence, both Miss Ossine and the girl Zwelferine. I took the precaution of telling the public of the reason for this arrangement, stating that these persons were in my home, so that their condition could be ascertained at any moment, and thereby would lend support to the truth. They remained there eight months after my departure from Vienna and only left on orders from a higher authority.

On arriving at Paris in February, 1778, I began to enjoy the rest and relaxation afforded there in the interesting company of the scientists and physicians of that capital. However, agreeing to their requests, and to repay the kindness shown to me, I decided to satisfy their curiosity by speaking of my system. They were astonished at its nature, and the results obtained, and asked me for an explanation. I gave them my concise assertions in nineteen articles. These assertions seemed to them to bear no relation to their established knowledge. I understood how difficult it was, by reason alone, to prove the existence of a principle of which people had not the slightest conception. With this in mind, I yielded to the request made to show the reality and the usefulness of my theory by the treatment of a few serious maladies.

A number of patients placed their trust in me. Most were in so desperate a plight that it required all my desire to be of use to make me decide on attending them. Nevertheless, I cured a vaporous melancholia with spasmodic vomiting, a number of long-standing stoppages in the spleen, liver and mesentery, an imperfect amaurosis, to the extent of preventing the patient from moving about alone, and a general paralysis with trembling which gave the patient (aged forty) every appearance of old age and drunkenness. This malady was the result of frostbite; it had been aggravated by the effects of a putrid and malignant fever which the patient had contracted six years before in America.

I also obtained the same success in a case of total paralysis of the legs, with atrophy; one of chronic vomiting, which reduced the patient to a state of progressive emaciation; one of general scrofulous debility, and finally, a case of general decay of the organs of perspiration.

These patients, whose condition was known and verified by the doctors of the Paris Faculty, were all subject to considerable crises, on a par with the nature of their maladies, without making use of any medicaments. After completing their treatment, they gave me detailed declarations.

It should have been more than enough to prove beyond doubt the advantages of my method. I had reason to flatter myself that recognition would follow. However, the persons who had asked me to undertake the foregoing treatments were not enabled to see their effects, owing to considerations and reasons which it would be out of place to enumerate in this dissertation. The result is that the cures which, contrary to my expectation, were not communicated to groups, whose duty it might have been to call the attention of the public to them, only imperfectly fulfilled the task I had set myself, and for which I had been praised.

This causes me to make a fresh effort today in the cause of truth, by giving more scope and the publicity which they have hitherto lacked to my original assertions.

The Philosophical Assertions of Franz Anton Mesmer

1. There exists an allied and reciprocal influence between the Heavenly Bodies, the Earth and those Bodies which are animated.
2. A universal fluid which is so continuous as to be without vacuum, and of a subtlety incomparable, which by its essence is capable of receiving, propagating and communicating all the impressions of movement — this is the vehicle of this influence.
3. This reciprocal force is subordinated to certain mechanical laws that were unknown until now.
4. This reciprocal action results in an alternation of cause and effect which may be regarded as a Flux and Reflux.
5. The Flux and Reflux are more or less general, more or less particular, and more or less composite, according to the nature of the causes that determines them.
6. By this operation, the most universal in all of Nature, active relationships are set between the heavenly bodies, the earth and all of the constituent parts thereof.
7. The properties of Matter and of the organized Body depend on this influence and its operation.
8. The animal body experiences the alternate effects of this agent. By insinuating itself into the substance of the nerves, it affects them at once.
9. There are properties in the human body that are similar to those of a magnet; the existent opposite poles may also be distinguished and these can be excited, changed, destroyed or reinforced, and even the properties of attraction and repulsion can be observed in the result.
10. This property of the animal body, which brings it under the influence of the heavenly bodies and the reciprocal action of those surrounding it, has induced me to call the force, Animal Magnetism, from its analogy with the Magnet,
11. The action and properties of Animal Magnetism, thus characterized, may be communicated to other bodies, both animate and inanimate, as these bodies are susceptible to it.

12. This action and properties may be propagated and strengthened by the same bodies.
13. The flow of a matter, subtle and rare, penetrates all bodies without a loss in activity. This may be observed experimentally.
14. The action and power of Animal Magnetism can be expressed from a great distance and requires no intermediary body.
15. It can be augmented and reflected by and through mirrors like light.
16. It may be communicated, propagated and augmented by sound.
17. This magnetic virtue, or property, may be accumulated, concentrated and transported.
18. I have said that all animate bodies are not equally susceptible. There are some few (a very few) who have opposite properties. The mere presence of these bodies destroys all the effects of magnetism in any other bodies.
19. This contrary property also interpenetrates all forms; it may likewise be communicated, propagated, stored, accumulated and transported, reflected by mirrors, and propagated by sound. This force then constitutes not merely a privation of magnetism, but a positive opposing influence.
20. The magnet, whether natural or artificial, is also susceptible to Animal Magnetism, and even to the opposing property, without its effect on iron and the needle undergoing any alteration. This proves that the principle of Animal Magnetism is different in its essential nature from magnetism of the mineral type.
21. This system will throw new light on the nature of Fire, Light and Heat, as well as on the theory of attraction, flux and reflux, the magnet, and of electricity.
22. It will show that the magnet and artificial electricity have an effect on maladies similar to other agents provided by Nature, and if some useful effects have come from their employment, these results are due to Animal Magnetism.
23. It will be recognized from the facts, according to rules I will establish, that this principle will cure all diseases of the nerves directly, and mediate all other diseases.
24. With its assistance, the physician is instructed and enlightened in the use of medicines. Indeed, he can hasten and improve their activity, and can bring about and direct beneficent crises, and make himself their master.

25. In communicating my method, I am propounding a new theory of illness by the universal utility of the principle I bring to bear.

26. With this knowledge, the physician will determine with certainty the origin, nature and progress of illnesses, even to the most complicated. He will prevent their advance and will succeed in curing them without exposing the patient to dangerous effects or unfortunate consequences, whatever be his age, temperament and sex. Even women in pregnancy and childbirth will enjoy these same benefits.

27. Finally, this doctrine will place the physician into a position to judge rightly the condition of each individual's health and safeguard him from maladies to which he might be exposed. The healing art will, by this doctrine and practice, attain to perfection.

My experience over a period of twelve years leaves me without doubt as to the efficacy and truth of these Assertions.* I realize that compared with old and established principles, my system appears to contain as much illusion as truth. I must, however, ask the enlightened to discard prejudice, or at least suspend judgment, until circumstance allows me to furnish the necessary evidence of these principles. In consideration for those languishing in pain and unhappiness due to the inadequacy of known methods, it is hoped this appeal will inspire the desire for more useful methods.

Physicians, being the trustees for everything connected with the preservation and happiness of mankind, are alone enabled, by the knowledge on which their profession is rounded, to judge the importance of this discovery which I have just announced and, moreover, to realize its implications. In short, they alone are qualified to put this method to practice.

As I have the privilege of sharing the dignity of their profession, I harbor no doubt that they will move to accept and spread those principles intended to alleviate the sufferings of Humanity, as they come to realize the importance of this Dissertation, which was written in essence for them. It is a true conception of ANIMAL MAGNETISM.

* Dr. Deslon, a physicisan of the Court, introduced these Assertions before a meeting of the Royal Society of Medicine — JB]

FOR A COMPLETE LIST OF PUBLICATIONS,
PLEASE ADDRESS:
HOLMES PUBLISHING GROUP LLC
POSTAL BOX 2370
SEQUIM WA 98382 USA
SECURE ORDERS:
JDHOLMES.COM